Tracking Your Carbon Footprint

Tracking Your Carbon Footprint

✦

A Step-by-Step Guide to Understanding and Inventorying Greenhouse Gas Emissions

Judith R. Purman

iUniverse, Inc.

New York Bloomington Shanghai

Tracking Your Carbon Footprint
A Step-by-Step Guide to Understanding and Inventorying Greenhouse Gas Emissions

iUniverse books may be ordered through booksellers or by contacting:

iUniverse
1663 Liberty Drive
Bloomington, IN 47403
www.iuniverse.com
1-800-Authors (1-800-288-4677)

Because of the dynamic nature of the Internet, any Web addresses or links contained in this book may have changed since publication and may no longer be valid.

The views expressed in this work are solely those of the author and do not necessarily reflect the views of the publisher, and the publisher hereby disclaims any responsibility for them.

ISBN: 978-0-595-50141-0 (pbk)
ISBN: 978-0-595-61389-2 (ebk)

Printed in the United States of America

to Paul
thanks for being my rock

Contents

Acknowledgments

Special thanks to Mr. Dan O'Neill for planting the seed, to Dr. Richard Ney for technical expertise and manuscript review, to editors Mr. Steven M. Maass, Ms. Joan Healy, and Mr. Doug Maust, and to Ms. Robin Tillman for the graphics.

Climate change has profound implications for virtually all aspects of human well-being, from jobs and health to food, and security and peace within and among nations.... Doom and gloom scenarios meant to shock people into action often end up having the opposite effect, and so it has been at times with climate change. We must focus not only on the perils but also on the associated opportunities.... People are yearning to do what it takes to address this threat and move to a safer and sounder model of development.

—Kofi Annan
November 2006

Preface

The world's attention has become increasingly focused on the issue and challenges of global climate change. The recent surge in energy prices and the ensuing shocks felt throughout the economy have only added to the urgency for action to reduce our dependence on carbon-emitting fossil fuels. Grand pronouncements of pending environmental doom, political intrigue that obscures the true picture, and competing economic interests place the issue of addressing climate change among the most volatile and important challenges of our time.

The good news is that this explosion of interest has spawned an equally impressive deluge of information about climate change, its potential impacts, and prospects for reducing those impacts. You now see climate-related information at almost every turn. The bad news is that this deluge of information comes from passionate, yet often competing sources, leading to a tremendous amount of misinformation and confusion for the public at large. Being a complex scientific issue at heart, much of the climate change literature is loaded with figures, terms and concepts that either confuse or quickly turn off the casual reader.

Judith Purman, with her simple yet elegant writing style has succeeded in crafting this brief manual to bring clarity to the process of measuring our impacts today, and managing those impacts to build a better future. She begins with a quiz to let the reader assess their general knowledge of climate change issues, and then works her way through creating the carbon footprint, setting emission reduction targets, developing management plans, and navigating the world of emission credit trading. Finally, Ms. Purman provides a concise list of resources from those key programs that have developed to take leading roles in the current voluntary marketplace for greenhouse gas emissions.

Rather than a lengthy, science-laden manuscript, this brief guide creates a path toward identifying and building a greenhouse gas management strategy, all in an easily read and understood format.

Richard Ney, PhD
Cedar Rapids, Iowa
April, 2008

Introduction

Whether you are a business, an organization or a household, knowing the extent of your greenhouse gas (GHG) emissions has become a hot button issue today. Why? Because more and more, consumers expect their employers, government, and schools to embrace the notion that one's style of living can negatively affect the environment today and for future generations. Likewise, homeowners, businesses, and organizations are moving to more sustainable modes of operating, not just because it *is* the right thing to do, but because sustainability, being "green," and reducing your carbon footprint have value in the marketplace. Sustainability is marketable and bankable, whether in dollars saved, in revenues generated, or in public relations impact.

How do you apply this practically? You clearly understand the importance and significance of getting it right but are uncertain of how to proceed. Perhaps you have committed to being "carbon neutral" but do not know exactly what that means. Perhaps you aren't sure how to put a plan in place to make changes that reduce your emissions or even what changes in your day-to-day lifestyle or business operations will result in emission reductions or avoidance. Perhaps you've heard about carbon credits and "cap and trade" and want to understand how this system can work for you.

Tracking Your Carbon Footprint: A Step-by-Step Guide to Understanding and Inventorying Greenhouse Gas Emissions has been designed to answer your questions. *Tracking Your Carbon Footprint* will introduce you to:

The basics of global climate change, global warming and GHGs
The what, why and how of GHG inventorying
How to use your inventory to set goals and reduce your GHG emis-

sions

How to determine whether or not you can generate carbon credits

How much do you already understand about these issues?

1

Establishing Your Baseline: How much do you already understand about climate change and greenhouse gas inventories?

As you will learn in Chapter 3, a greenhouse gas inventory begins with establishing a baseline of emissions. You cannot plan to reduce emissions unless you understand what you currently emit. Similarly, you cannot understand the process of inventorying GHGs until you have baseline knowledge of GHGs themselves, their effect on climate change, and the process of how to inventory.

Test yourself. What's your baseline knowledge of GHGs and GHG inventories?

What's your baseline knowledge?

1. Weather includes

    a. average annual temperatures

    b. change of seasons

    c. precipitation patterns

    d. daily temperatures, cloud cover and precipitation

2. Climate change is caused by

    a. natural changes in the environment

    b.   burning fossil fuels

    c.   land filling waste

    d.   all of the above

3. Global warming

    a.   is a rise in average annual temperatures over time

    b.   results in a decrease in average annual temperatures over time

    c.   is one manifestation of global climate change

    d.   all of the above

4. The greenhouse effect

    a.   makes it possible for the earth to sustain life

    b.   would not exist without fossil fuel use

    c.   causes infrared radiation

    d.   induces sun spots

5. A greenhouse gas is a gas that:

    a.   cools as it hits the earth

    b.   reflects heat in the atmosphere

    c.   prevents cool air from reaching the earth

    d.   absorbs infrared radiation reflected from the earth's surface

6. The three main greenhouse gases related to global climate change are:

    a.   Nitrous oxide, carbon dioxide, and sulfur

    b.   Carbon dioxide, nitrogen, and helium

    c.   Methane, radon, and carbon dioxide

    d.   Carbon dioxide, methane and nitrous oxide

7. True or False: All GHGs have equal effect on the environment.

8. Global warming potential is

   a. a means of comparing the affect of GHGs on global climate change

   b. a means of measuring how a GHG will affect global climate change over 100 years

   c. a standard of comparison for gases using carbon dioxide as a base

   d. all of the above

9. If carbon dioxide has a global warming potential of 1 and methane has a global warming potential of 23, and the concentration in the atmosphere is the same for both,

   a. carbon dioxide will have a greater affect on global climate change

   b. methane will have a greater affect on global climate change

   c. carbon dioxide and methane will equally affect global climate change

   d. none of the above

10. "Carbon footprint" refers to

   a. the measure of all GHGs emitted by a defined entity due to the burning of fossil fuels

   b. the measure of carbon dioxide emitted by a defined entity due to electricity use

   c. the measure of methane emitted by a defined entity

   d. the measure of all GHGs emitted by a defined entity

11. One way to lessen your carbon footprint is

   a. to drive rather than fly

   b. to walk rather than drive

    c.   to use electricity generated by wind rather than coal-generated electricity

    d.   all of the above

12. Quantifying the amount of GHGs emitted includes

    a.   tracking emissions over time

    b.   defining a starting point

    c.   calculating the emissions in the same way over time

    d.   all of the above

13. A GHG inventory is a(n)

    a.   listing of the source of GHG

    b.   accounting of the quantity of GHGs emitted, removed or sequestered over time

    c.   necessary component of global climate change

    d.   system of understanding GHGs used only by businesses

14. Of the following, what is NOT included in a GHG inventory?

    a.   methods used to track data

    b.   rationalization for why certain portions of your emissions were excluded from the count

    c.   description of emission effects on local weather

    d.   base year determination

15. GHG inventories are conducted by

    a.   governments

    b.   businesses, schools, and nonprofits

    c.   households

    d.   all of the above

16. Reducing GHG emissions through sequestration means
    a. preserving carbon in storage
    b. not emitting carbon
    c. land filling food waste
    d. none of the above

17. Reducing GHG emissions through avoidance means
    a. turning off the lights when you leave the room
    b. composting food scraps rather than land filling them
    c. walking rather than driving
    d. all of the above

18. Carbon credits
    a. are a function of a market system that gives value to the credit
    b. result from reducing or avoiding GHG emissions or sequestering carbon
    c. are generated by documented emissions reductions
    d. all of the above

19. Carbon credits are issued by
    a. the New York Stock Exchange
    b. the federal government
    c. voluntary exchanges where credits are bought and sold
    d. all of the above

20. Establishing an inventory boundary is
    a. a way to exclude factors which make your emissions high
    b. important for establishing legitimacy of your inventory

    c.   vital for documentation of changes to emissions

    d.   a and b

    e.   a and c

    f.   b and c

21. True or False: If you are a small business, your inventory boundary may include the fuel your employees use to commute to work and the emissions from the power company that supplies your office building with electricity.

22. Considerations for choosing a base year for your inventory include:

    a.   the earliest point in time for which relevant data is available

    b.   the exchange through which you are hoping to sell carbon credits

    c.   what the local regulations may require as a base year

    d.   all of the above

23. True or False: You can pick any base year from which to compare future emissions reductions as long as you justify your choice.

24. True or False: Emissions can be calculated using either actual data or emissions factors applied to your data.

25. Reducing GHG emissions through reduction, avoidance or sequestration can generate carbon credits if you

    a.   sell those reductions on one of the established exchanges

    b.   meet the conditions of all of your permits

    c.   conduct an inventory

    d.   have the reductions certified by a third party verifier associated with the exchange of which you are a member

26. Emissions trading includes

    a. buying and selling of carbon credits

    b. conducting GHG inventories

    c. cap and trade systems

    d. none of the above

27. A "cap and trade" system

    a. has been shown to achieve environmental goals more quickly than imposing regulation

    b. is a type of regulation

    c. uses the marketplace to find the lowest cost path to compliance

    d. all of the above

28. An "offset project"

    a. is generally undertaken by non-regulated entities

    b. is funded by the offset purchaser

    c. results in reductions in GHG emissions with the reductions sold on an exchange

    d. all of the above

29. An "offset project" may include

    a. changing light bulbs

    b. riding your bike to the store instead of driving

    c. improving the efficiency in a factory

    d. all of the above

30. True or False: The Kyoto Protocol is a carbon trading exchange.

How did you do?
See Appendix A to check your answers.

2

Climate Change Primer

If you completed the quiz in Chapter 1, you now have a clearer picture of what you do and do not understand about GHGs and GHG inventories. Study of the Climate Change Primer will help to clarify for you the basic science of climate change and the terminology used to discuss GHGs, climate change, and GHG inventories.

What is the difference between weather and climate?

Weather is what happens every day, in a given place, at a given time, with respect to heat or cold, wetness or dryness, calm or storm, clearness or cloudiness. Climate, on the other hand, describes the total of all weather occurring over a period of years in a given place. Climate includes the change of seasons, special weather events like tornadoes and floods, and precipitation patterns.

What is climate change?

Climate change refers to any significant change in climate indicators. Climate indicators include such things as precipitation, temperature, and storm patterns and intensity. Climate change can result from natural factors and human activities.

We know that there were ice ages and periods of global warming before humans inhabited the earth. Scientists do not know definitively why these changes have occurred, but they have theories about their causes. For example, volcanic eruptions on a massive scale could produce enough smoke, particulate matter and extra gases to shade the earth and prevent sunlight from passing through the atmosphere. This would result in a decrease in temperature. Any changes in the earth's orbit or the earth's orientation toward the sun would affect the earth's temperature. "Sun spots," explosions on the sun, would increase the amount of heat generated by the sun, which, in turn, would raise the temperature on the earth's surface.

Human activities known to affect the climate include burning fossil fuels, deforestation, desertification, land filling and combustion of waste, and agricultural practices such as soil cultivation, rice production, and animal husbandry.

What is global warming?

Global warming refers to an *average* increase in the earth's temperature over time. Global warming is one manifestation of climate change. It is important to realize that warming trends are not uniform throughout the globe. Increases in the average temperature in one region of the world may produce far different results in another region.

What is a Greenhouse Gas (GHG)?

A greenhouse gas is any gas that absorbs the sun's radiation in the atmosphere. Some GHGs occur naturally and are emitted to the atmosphere through natural processes. Other GHGs result from human activities such as the use of fossil fuels, deforestation, and some methods of farming.

Which GHGs are important to global warming and climate change?

There are six main GHGs that contribute to global warming: carbon dioxide, methane, nitrous oxide, hydrofluorocarbons, perfluorocarbons, and sulphur hexafluoride. Table 1 outlines the chemical symbol and sources for each of these.

TABLE 1

SIX GREENHOUSE GASES AND THEIR SOURCES

GHG	SYMBOL	SOURCE
Carbon Dioxide	CO_2	Burning of fossil fuels, solid waste, trees and wood products. Chemical reactions.
Methane	CH_4	Coal, natural gas production and transport. Livestock operations. Decay of organic waste in landfills.
Nitrous Oxide	N_2O	Agricultural and industrial activities. Combustion of fossil fuels and solid waste.
Hydrofluorocarbons	HFC_s	Refrigeration
Perfluorocarbons	PFC_s	Electronics manufacturing
Sulphur hexafluoride	SF_6	Electric transformers

What is the greenhouse effect?

The greenhouse effect is a natural process that maintains the earth's temperature at about 60°F above what it would be without the greenhouse effect. It is fundamental to life's existence on earth. Without the greenhouse effect, the life systems as we know them would not exist.

It works like this. Every day, sunlight passes through the atmosphere to the earth's surface where it is converted into infrared radiation (heat). Some of the heat is absorbed by the earth's surface, some bounces off the earth's surface and escapes into space, and some radiates back from the earth's surface into the atmosphere where it is trapped by greenhouse gases and re-emitted to the earth's surface. Gases that trap heat in the atmosphere are known as greenhouse gases (GHGs). See Figure 1.

Without the greenhouse effect, the earth could not sustain life. If this effect becomes stronger, the earth becomes warmer. This is "global warming". An increase in temperature of only a few degrees can lead to dramatic consequences for humans, plants and animals.

FIGURE 1
THE GREENHOUSE EFFECT

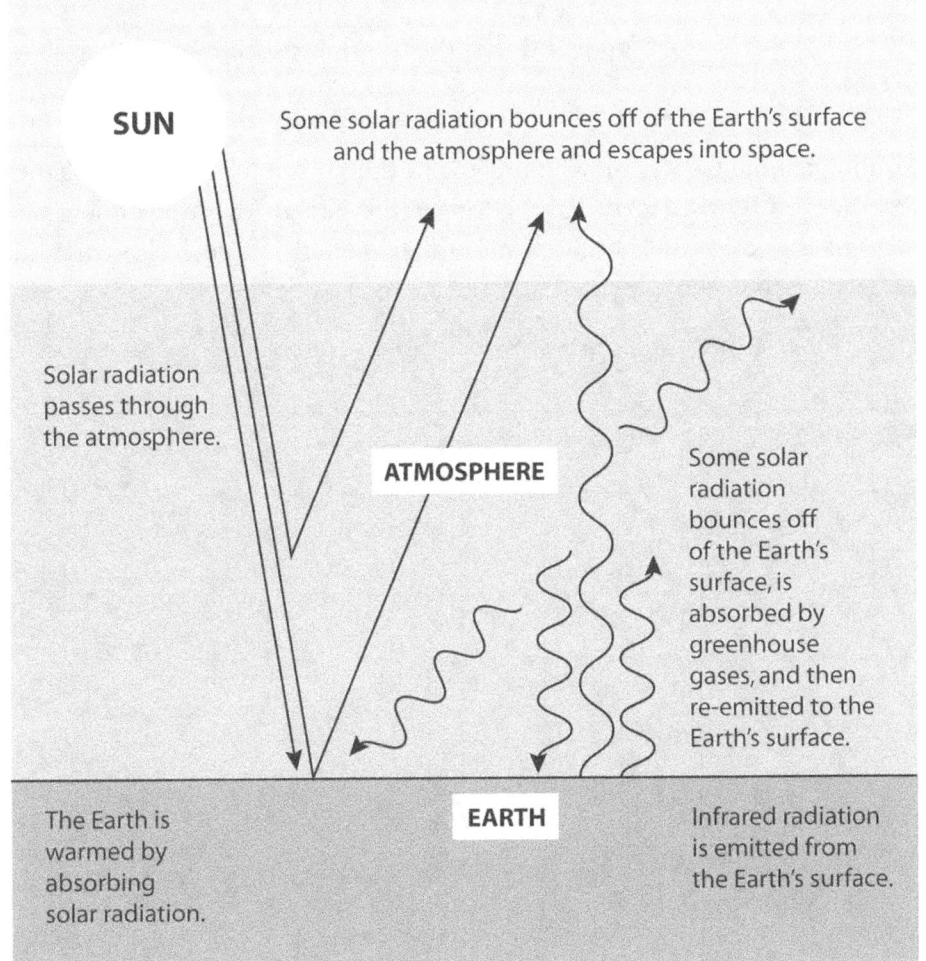

What is global warming potential?

Global Warming Potential (GWP) is a measure of how much a gas contributes to global warming over a period of time (100 years) compared to carbon dioxide. Carbon dioxide has been assigned a GWP of 1 since it is the most prevalent GHG. GWP allows you to compare the impact of the concentrations of the GHGs to each other and carbon dioxide. For example:

Example: CH_4 has a GWP of 23
                     1 kilogram (kg) CH_4 = the emission impact of 23 kg
of CO_2

Although some GHGs may be present in the atmosphere in lesser quantities than CO_2, they may have a longer lifespan in the atmosphere and may, in the long run, be much more detrimental. Table 2 compares the GWP of carbon dioxide, methane and nitrous oxide.

TABLE 2

COMPARISON OF THE CONCENTRATION
AND GWP OF 3 MAIN GREENHOUSE GASES

	CO_2	CH_4	N_2O
Concentration in atmosphere	ppm	ppb	ppb
Pre- industrial	280	700	270
Current	370	1745	314
Lifetime in atmosphere (years)	5-200	12	114
CO2 equivalent or GWP	1	23	296

Source: Intergovernmental Panel on Climate Change, Third Assessment Report.
* Note: ppm = parts per million, ppb = parts per billion

Several interesting conclusions are evident from Table 2. Methane remains in the atmosphere the shortest period of time. Reducing 1 kilogram (kg) CH_4 will have the same effect on global warming as reducing 23 kg of CO_2. Nitrous Oxide is more harmful than either carbon dioxide or methane.

This becomes important in making decisions about emission reductions. Which GHG will you target to reduce and how will you make that decision? Can you more economically or easily reduce 1kg of CO_2 or 1 kilogram (kg) CH_4? Which has the greater impact on global warming?

What is meant by carbon footprint?

The term "carbon footprint" refers to the measure of all GHGs emitted by a defined entity. The carbon footprint of individuals and families may include the amount of GHGs emitted to run the household: the use of the air conditioner, the furnace, the washer and dryer, the car, etc. For organizations or businesses, the carbon footprint may include the amount of GHGs emitted as part of normal operations: burning fuel to power machinery, disposal of waste materials including electronics, employee commuting, etc. The carbon footprint for cities, states and countries may include GHGs emitted by transportation systems, power generating facilities, industry, agriculture and manufacturing.

How do you quantify your emissions?

You quantify your emissions by conducting a GHG emission inventory based on sound design principles!

Key Points

- Climate change and global warming are not the same thing.
- Greenhouse gases affect the environment differently depending on concentration and how long they persist in the atmosphere.
- The GWP (Global Warming Potential) is a measure used to compare the effect of one GHG on the environment over 100 years compared to carbon dioxide.
- The carbon footprint accounts for all GHG emissions generated by a defined entity. Carbon footprints can be calculated for homes, businesses, schools, cities and countries.

3

GHG Inventories: Determine Your Carbon Footprint

You now understand the basic science and terminology related to GHGs. Now you are prepared to begin the process of quantifying your carbon footprint in a GHG inventory.

What is a GHG Inventory?

A GHG inventory is a systematic accounting of existing GHG emissions for a defined entity over a given period of time. Inventories can be undertaken on any level: global, national, state, local, by company, educational institution, or by household. In an inventory, you will identify, calculate, verify and report your emissions.

What is included in an inventory?

Your inventory should include:

- A defined entity you want to inventory

- A listing of the GHGs you plan to track, along with your calculation methods

- Base year standards to serve as a beginning point

- Documentation of all your activities that emit or remove those GHGs

- Methods you plan to use to account for quantities emitted or removed

A number of organizations provide guidelines and recommendations on how to conduct inventories. The International Panel on Climate Change (IPCC) guidelines are for nation-level inventories such as those participating in the Kyoto Protocol. The Environmental Protection Agency's (EPA) Climate Leaders Program and the Pew Center on Global Climate Change also have a set of guidelines geared toward large corporations. All of these guidelines are based on the same inventory design principles.

What are Inventory Design Principles?

Thinking carefully about how you will conduct your inventory is important for gaining the information you seek. In *The Greenhouse Gas Protocol, A Corporate Accounting and Reporting Standard,* the World Resources Institute (WRI) recommends the five principles around which to frame your inventory design (Figure 2)

Relevance: You want your inventory to reflect the emissions within the defined boundary, whether the boundary includes a household, one factory, or a national company. You also want your inventory to serve the decision-making needs of those using the inventory.

For example, if you are inventorying the emissions of your household, you are the one who will use the findings of the inventory to reduce your emissions. Think about what you need to *quantify* in order to make an informed decision about lifestyle changes. If you run a business or organization, determine who will ultimately make the decisions on changes in business practices identified as important by the inventory. It will be helpful to consult those decision-makers throughout the inventorying process to ensure from the outset that you have designed the inventory to provide the data they need to make informed decisions.

Completeness: You want your inventory to include all GHG emission sources and activities within your chosen inventory boundary.

For example, if you run a factory but decide not to include the fuel used to run your front end loaders and forklifts, you will have to explain and justify that exclusion. Or, if your organization has three separate offices in the same city and you decide to exclude one of those offices from your inventory, you will have to justify that exclusion. Does it make sense not to count that equipment or that office?

Think carefully about how you choose your boundary; what to include and what to exclude. This is very important if your inventory is to be verified. Verification is a requirement when you are generating carbon credits or aiming for certification.

Consistency: You want to be consistent in how you gather data and estimate emissions so that you can compare emissions over time.

For example, if you base your emission calculations for heating on past bills one year but the following year use emission factors instead, you have applied two different calculation methods from one year to the next. Do the two methods result in an apples-to-apples comparison? Why did you switch methods? (More on calculation methods in Step 3 below)

Transparency: Verification or your method of inventorying is required by the exchanges buying and selling carbon credits. If your goal is to sell carbon credits, it is very important that your methodology as well as data collection and estimation of emissions be clearly documented and available.

Accuracy: You want to be as accurate as possible in your data collection and emissions estimation.

Refer to the sidebar for more information on the World Resources Institute.

FIGURE 2

PRINCIPLES FOR GREENHOUSE GAS INVENTORY DESIGN

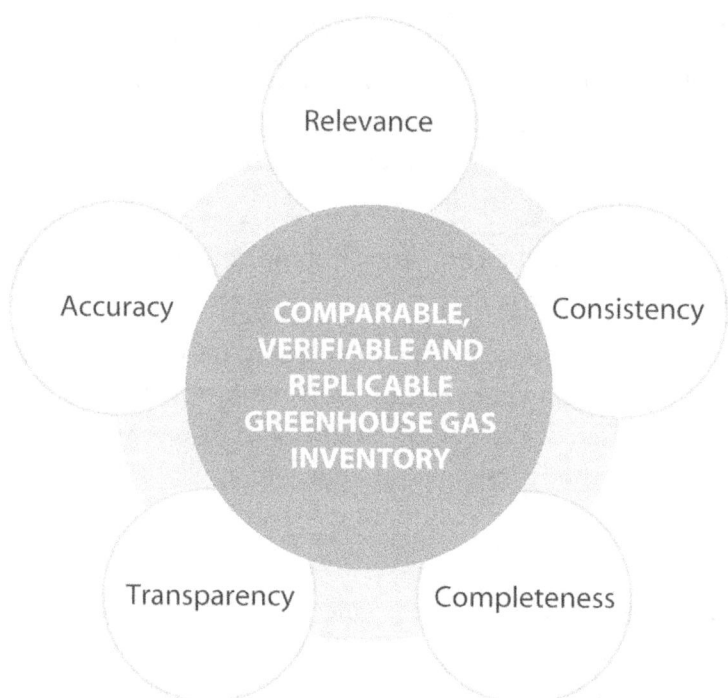

Why is it important to follow accepted design principles and protocols for conducting inventories?

If you are going to the trouble of conducting an inventory in the first place, you want to generate as accurate an estimation of your carbon footprint as possible. The design principles will guide your process. Following the standards and guidelines is important to ensure that GHG inventories are conducted in such a way that the results are comparable and understandable.

If you are planning to make emission reductions for the purpose of generating carbon credits to sell in a marketplace, verification by independent third parties is required.

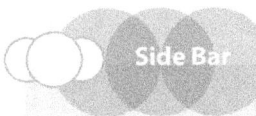 **Side Bar** **The World Resources Institute (WRI)**

The World Resources Institute (WRI) is an environmental think tank whose mission is to inspire and move human society to protect the Earth's environment and preserve its ability to provide for current and future generations. WRI is committed to "protect the global climate system from further harm due to emissions of greenhouse gases and help humanity and the natural world adapt to unavoidable climate change."

WRI has published the Greenhouse Gas Protocol Initiative, in conjunction with the World Business Council for Sustainable Development (WBCSD), to promote GHG inventorying and mitigation projects. Through this initiative, they have established internationally accepted standards for the process of inventorying and reporting emissions and mitigation projects.

For more information, go to WRI at *www.wri.org*.

FIGURE 3
STEPS TO A COMPREHENSIVE GHG INVENTORY

1. Determine Inventory Boundary

2. Select Base Year

3. Calculate Greenhouse Gas Emissions

4. Verify by Independent Third Party

What are the steps involved in inventorying GHGs? (Figure 3)

There are 4 basic steps in an inventory:

1. Determine Inventory Boundary

2. Select a Base Year

3. Calculate GHG emissions

4. Verify by independent third party (if you are seeking to trade carbon credits)

Each of these steps involves sub-steps. Now is the time to begin applying the inventory design principles by walking step-by-step through the nitty-gritty detail of conducting your inventory.

Step 1: Determine Inventory Boundary (Figure 3a):

Setting reasonable boundaries for the inventory is your first step. How will you consolidate your GHG emissions data for accounting and reporting purposes? What should you include in your inventory?

For example, if you are a household and your power company burns coal to generate the electricity you use, should you count the emissions of the power company generating your electricity as well the coal mining operation and transportation of the coal to the power plant? If you purchase food from the grocery store, should you include your gas to the grocery store as well as the emissions from the trucks transporting the food from the processing plant to the store? These can be daunting questions. Where does your responsibility start and stop? What is within your control? A well defined inventory boundary helps you to clarify what should and should not be included.

By definition, the inventory boundary includes both organizational and operational processes.

Inventory Boundary = organizational boundary + operational boundary

A. **Set organizational boundary**:

Choose one of two possible approaches:

1. **Equity share approach**: This is GHG accounting based on your economic interest in the organization. What percent ownership do you have in that operation?

   If you own half of Company X and half of Company Y, under the equity share approach to establishing an inventory boundary, you will include half of Company X's and half of Company Y's emissions in your inventory.

   If you and one partner are equal owners of a trucking company, you own 50% of the economic interest in that company and will accept 50% of the responsibility for emissions inventoried.

2. **Control approach**: This is GHG accounting based on 100% of the emissions from operations over which you have control. If you "have control" over operations, this means you have the ability to change operations within that defined boundary. For example, a homeowner can turn down the thermostat or delivery trucks could be fueled on biodiesel instead of gasoline. These would be included under a control approach to boundary definition. On the other hand, emissions from the power company from which you purchase electricity may not be included because you, as their customer, cannot change the power company's operations.

B. **Set operational boundary:**

To determine your operational boundary, you must:

1. Identify the operations that create emissions.

2. Identify emissions associated with those operations.

3. Categorize those emissions as direct or indirect. A direct emission is one that you control or own. Emissions from the furnace you run to heat your home or office is an example of a direct emission.

An indirect emission is generated by sources controlled by another company. Electricity is an example of an indirect emission; you, the homeowner or business, use electricity but you do not generate electricity.

FIGURE 3a

1. Determine Inventory Boundary

A. Set Organizational Boundary

　　1. Equity Share Approach

　　2. Control Approach

B. Set Operational Boundary

　　1. Identify Operations

　　2. Identify Emissions

　　3. Categorize Emissions

Step 2: Select a Base Year (Figure 3b)

The base year is the year to which you will compare your current and future emissions. This is important because it establishes a baseline for future changes in your emissions.

Existing emissions trading programs and registries require a fixed base year structure. The Kyoto Protocol established 1990 as the base year to which to compare emissions. Refer to the side bar for further information on the Kyoto Protocol.

Different registries have in place a variety of rules and regulations about base years. It is becoming increasingly likely that registries, markets and regulations will become the norm. Regardless of which system is in place, the base year will be determined by that system. It is important, if you are an organization or corporation, for you to know what the rules and regulations are as you determine what your base year will be.

For now, you may want to choose a year based on the earliest point in time for which you have relevant data. For example, if you are a homeowner who has been in your home for 2 years, you would consider using gas and electricity data for the first full year you were in your home.

Pick a base year and provide justification for your choice.

FIGURE 3b

2. Select Base Year

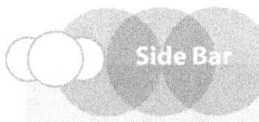

Side Bar   **The Kyoto Protocol**

The Kyoto Protocol is a United Nations sponsored agreement between nations to reduce GHG emissions. This voluntary agreement has been ratified by over 140 countries. Specifically, it relates to GHG emissions levels for the years 2008-2012 as compared to 1990 levels (the base year). The Protocol is the 1997 amendment to the 1992 United Nations Framework on Climate Change. This treaty is the first of its kind and may have wide implications in the transfer of national powers to an international governing body. This is unprecedented in its potential to dictate energy policy to those countries that ratified it.

Under the terms of the agreement, industrialized countries have individual emissions targets that add up to a total cut in GHG emissions of at least 5% from 1990 levels by 2012. Specifically, the European Union nations and Canada committed to an 8% reduction, while Australia is allowed an 8% increase. The Protocol does not commit poor or developing nations including China, India or Mexico. This agreement is legally binding upon the signatories.

The United States theoretically committed to a 7% reduction but has yet to ratify the Protocol. Even though the United States has not ratified the Protocol, many multinational corporations embrace and operate under its tenets. In addition to the multinational corporations, the United States Mayors Climate Protection Agreement was approved by the United States Conference of Mayors in June 2005 and now has the endorsement of over 220 mayors. Under this Agreement, mayors pledge to meet or exceed the Kyoto targets for their city as well as to work to develop, enact and promote local, state and federal initiatives which support the Kyoto Protocol.

For more information, go to *www.ipcc.ch* or *www.usmayors.org*.

Step 3: Calculate GHG Emissions (Figure 3c)

Now you are into the real meat and potatoes of the inventory. Step 3 includes identification of emission sources and calculation of actual emissions.

Two types of GHG protocol tools can assist you in determining your emissions calculations. These include cross-sector and sector-specific approaches. Most inventories will use both types depending on which emission sources are to be included. These tools are available on www.ghgprotocol.org, www.epa.gov/climateleaders, www.iclei.org, and www.cleanair-coolplanet.org.

A. **Identify emission sources:**

1. Categorize GHG sources within your defined boundary. These may include:

   a. Fuel combustion in stationary equipment (furnaces, engines, flares, etc.)

   b. Fuel combustion in mobile equipment (cars, trucks, heavy equipment, planes, etc.)

   c. Process emissions (chemical processes in manufacturing, petrochemical processing, etc.)

   d. Fugitive emissions (wastewater treatment, land filling, etc.)

2. Identify the GHG sources as direct or indirect emissions as defined in Step 1 B(3).

B. **Select a calculation approach:** Will you measure emissions directly or calculate emissions based on emission factors? You should use the most accurate approach possible. Refer to the side bar for further information on the use of online emissions calculators.

Measuring emissions directly means that you have the on site equipment necessary to collect air samples and analyze them for GHG emissions. For example, emissions from heavy equipment in a factory can be measured by specialized air sampling equipment placed in the factory. The equipment samples the air on a regular basis for a designated period of time and then analyzes the samples. The results from the sampling and analysis provide the emissions data.

An emission factor allows GHG emissions to be estimated from a unit of available data. Emission factors are used when direct monitoring is not available or is cost prohibitive. It works this way: you may know both the amount of fuel you used and have access to data on the carbon content of that fuel. These two combined can be used to estimate what the emissions were with reasonable accuracy.

Online Emissions Calculators

On line emissions calculators have been developed to estimate GHG emissions for households. You input the requested data, click Enter, and the calculator determines your carbon footprint for you. Some examples of these can be found on the following websites:

www.epa.gov/climatechange/emissions/ind_calculator.html
www.carbonify.com/carbon-calculator.htm
www.marama.org/diesel/calculators.html
www.cleanerandgreener.org/resources/pollutioncalculator.htm

This list is not meant to be all-inclusive. Many, many sites now include these types of calculators. While the data generated is interesting and provides enough information to change personal behavior, calculations of emissions by these on line calculators is not accepted as a legitimate determination for carbon credit generation.

C. **Collect data:** This is the nuts and bolts of your inventory. Consistency in collection and calculation is essential to ensuring accuracy! You've selected your boundary in Step 1, selected your base year in Step 2 and figured out what you will track and how you will calculate (using actual data or emission factors) identified emissions in Steps 3A and 3B. Now, pull together the data, research those emission factors and set up a spreadsheet to track the data. Consult someone experienced in inventorying to guide you and to review your work to ensure accuracy and completeness. This is especially important if you are interested in generating carbon credits.

FIGURE 3c

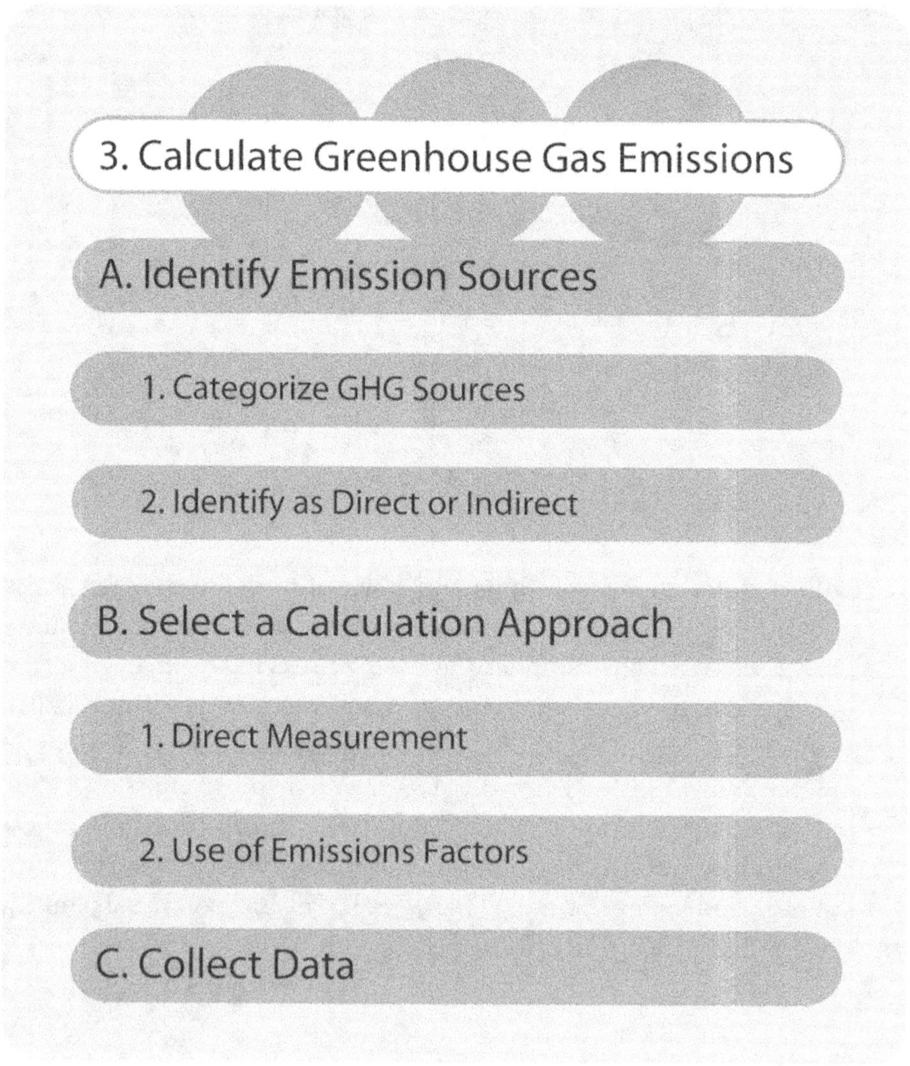

3. Calculate Greenhouse Gas Emissions

A. Identify Emission Sources

1. Categorize GHG Sources

2. Identify as Direct or Indirect

B. Select a Calculation Approach

1. Direct Measurement

2. Use of Emissions Factors

C. Collect Data

Step 4: Third Party Verification (Figure 3d)

If you are a business or organization interested in generating carbon credits, the exchange where you will sell your credits will hire an independent third party to assess how you conducted your inventory and the quality of your data. This is to establish that your inventory represents a true and fair account of your GHG emissions. Verification will include site visits and data review.

FIGURE 3d

4. Verify by Independent Third Party

Now that you've conducted your inventory, what's next?

You now understand which sources emit how much of specific GHGs. Your next step is to identify where in your overall system you can reduce and/or avoid emissions and sequester (store) carbon.

Key Points

- A GHG inventory is a systematic accounting of existing GHG emissions for a defined entity over a given period of time.
- GHG inventory methodology is based on five important principles: relevance, completeness, consistency, transparency and accuracy.
- The steps included in an inventory are boundary establishment, base year selection, emissions calculation, and verification. Figures 3a-d summarize these steps.
- Verification by an independent third party is essential if you are seeking to trade carbon credits.
- GHG inventories provide the information necessary to determine where and how you can lessen your carbon footprint.

4

Using Your Inventory To Set and Meet GHG Reduction Goals

You have now successfully completed the inventory of your GHG emissions for your defined entity. Your next step is to determine how you can interpret your inventory to set goals to reduce your carbon footprint. Are you able to reduce emissions? Avoid emissions? Sequester carbon?

What is the difference between emission reductions, avoidance and sequestration?

Reducing GHG emissions means that you have successfully lowered the amount of GHGs released. For example, if you drive the same number of miles as usual but you buy a more fuel efficient car, you have reduced your GHG emissions by the increase miles per gallon driven. If your factory switches from using electricity generated by a coal-fired power plant to electricity generated from wind power, you have lowered the amount of GHGs generated in the production of electricity you use in the factory.

Avoiding emissions means that you have changed your operations in such a way as to prevent GHG emissions that otherwise would have been generated. For example, if you walk to the store instead of driving, you have avoided those emissions you would have generated by running your car, or, if instead of throwing your food waste in the landfill, you opt to compost it in your backyard, you have avoided the generation of methane that amount of food waste would have produced in the landfill minus the amount of methane composting the material generated. (While composting itself generates some methane, it generates less methane than land filling the same material.)

Sequestration is preserving carbon in storage. The most significant sequestration occurs naturally as plants and trees secure carbon into the soil as they grow. Also, producing compost from food waste and yard waste captures carbon in the compost itself. When compost is applied to the soil, the carbon is sequestered in the soil.

How do you go about setting a goal to reduce your carbon footprint?

The process for setting goals to reduce GHG emissions mirrors the process for inventorying. See Figure 4. In this process, you will use the information in your inventory to target areas for change.

1. **Decide on a target type:** Does it make more sense to reduce emissions over time or to reduce the ratio of emissions relevant to a related entity? For example, if you are a homeowner, you may decide you want to reduce your CO_2 emissions by 10% from your base year. If you are a school, you may opt to reduce CO_2 emissions by 3% per student per year. This latter type of reduction goal is called an "intensity target."

2. **Decide on the target boundary:** Which GHGs can you include and which do you have little control over? Which GHGs will you include from which sources?

3. **Choose a base year:** Will you target reductions beginning with your base year or from last year? Remember to keep in mind your base year chosen for your inventory is determined by either the exchange you have joined or, if you have not joined an exchange, by the year for which you have the most complete data.

4. **Choose a targeted completion date:** What's a realistic timeframe for action and results? Will the actions be on-going or for a specified timeframe? These are important considerations as you decide what is sustainable for you in the long run.

5. **Track and report progress:** As with the inventory itself, it is crucial that you document the changes in emissions resulting from the changes you have made in how you operate. You must also use the same calculation method that you used in your inventory.

FIGURE 4

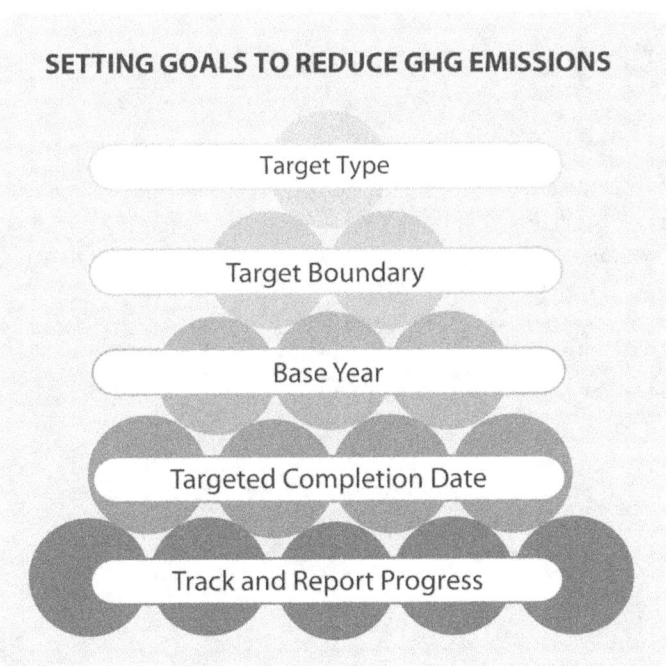

SETTING GOALS TO REDUCE GHG EMISSIONS

Target Type

Target Boundary

Base Year

Targeted Completion Date

Track and Report Progress

What is a reasonable goal for reducing your GHG impact?

Goal setting is entity-specific. "Reasonable" will depend on what your goals were at the outset. Were you interested in reducing as much as possible? Were you interested in reducing in one particular area? Were you interested in reducing enough to generate carbon credits? Were you interested in making annual changes to reduce emissions for the next 5 years?

Key Points

- Figuring out what is a realistic and achievable goal for reducing your carbon footprint is unique for each person, organization, company or government.
- Your inventory provides key information for analyzing what is a realistic GHG emission reduction goal for your defined entity.
- The process of setting a goal and implementing a plan to achieve that goal mimics the process of conducting your GHG inventory.

5

GHG Inventories, Carbon Credits, and Emissions Trading

You have conducted your GHG inventory, analyzed in detail what is contributing to your carbon footprint, designed and implemented a plan to reduce those emissions. Can you reduce, avoid or sequester enough to generate carbon credits?

What is a carbon credit?

A carbon credit is a value assigned to an entity's reduction, avoidance, or sequestration of GHG emissions. Carbon credits are a function of a market without which there would be no credits. Markets issue credits for reductions that are quantifiable, permanent, verified, and certified. The relationship between carbon credits and your GHG inventory is depicted in Figure 5.

FIGURE 5

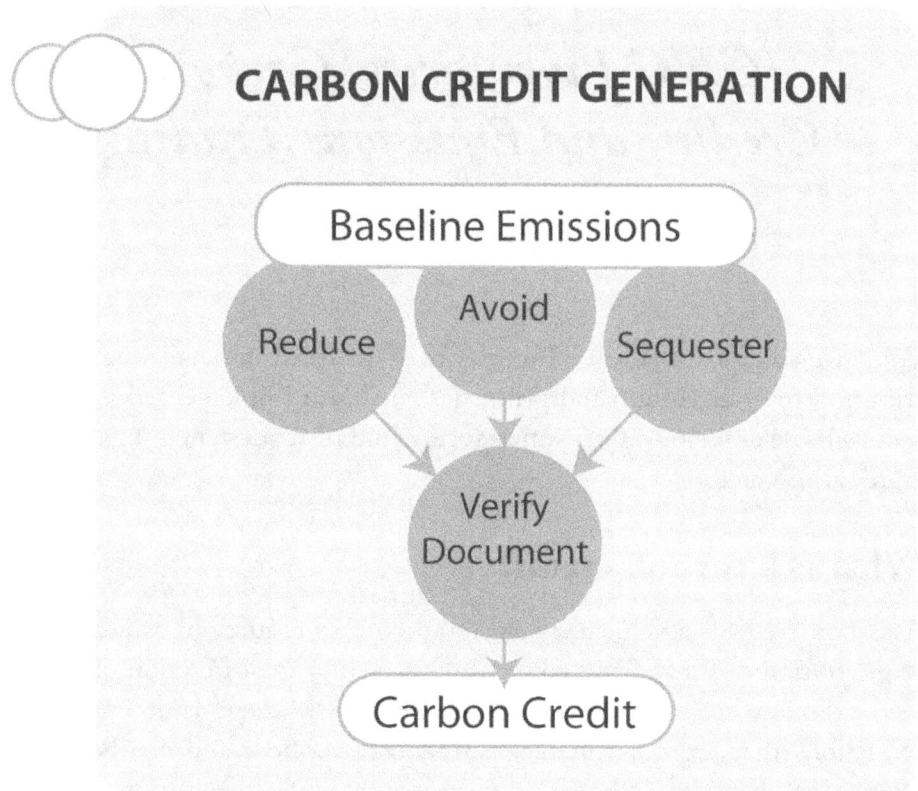

CARBON CREDIT GENERATION

Baseline Emissions

Reduce   Avoid   Sequester

Verify Document

Carbon Credit

How is a carbon credit generated?

Credits are issued for documented emission reductions or for documented changes in business practices that avoid GHG generation. You cannot be granted credit for an existing practice or system that is already established to reduce your emissions. Only new changes to your operations qualify for credit.

For example, if the cafeteria at your school has reduced the material it sends to the landfill by separating food waste for hog feed, switching to composting this material instead will not result in carbon credit generation because you were already managing this material. If your school had no system in place to recover food waste, the implementation of a food-to-hogs program or a composting program may result in credits for reduced or avoided emissions.

Who issues carbon credits?

Credits are issued according to standards unique to exchanges. Exchanges work somewhat like the New York Stock Exchange; only members can participate, credits are bought and sold based on market conditions. This is known as emissions trading. The price of a credit is affected by supply and demand. If very few credits are available, demand will be high and the price will rise. If many credits are generated, the price will drop.

What is emissions trading?

Emissions trading is a closed system, referred to as an "exchange," whereby credits for GHG emissions savings are bought and sold.

Under an emissions trading program, credits are granted for documented and verified reductions in GHG emissions. These credits are then sold by those that generated them, either directly or through a broker, and purchased by those who can purchase the credit more cheaply than they can generate their own reductions.

Members of the exchange pledge to abide by the rules of the exchange. The exchange limits or "caps" the total number of emission allowances for a particular GHG and for each member. Trading occurs when a member of the exchange has not used all of its allotted allowances. These "credits" are sold to members who require allowances because they generate more emissions than they are allowed under the rules of the exchange.

For example, Company A, after conducting a GHG inventory, determines that it can lower its carbon emissions by 10% by allowing half of its staff to work from home rather than commute to the office. After this operational change is implemented and the GHG emissions have been documented and verified, the GHG emissions saved generate a credit that Company A can sell to Company B whose own emission reductions proved to be cost-prohibitive.

For example, Country X reduces its net carbon emissions by planting trees. The trees sequester carbon dioxide. The amount of saved greenhouse gas generates a credit that Country X can sell to another country. The price is determined by mutual agreement. Country Y may choose to purchase these credits rather than institute measures to reduce its own emissions.

In both examples, a situation was created where actual emissions are reduced at the lowest possible cost. Company A and Country X generated credits through quantifiable and verifiable emission reductions. Company B and Country Y purchased the credits to meet the reduction quota required by the exchange. In this way, Company B and Country Y are credited for the reductions, not those who generated them.

What exchanges are currently operating?

Currently there is no federal system in place for buying and selling carbon credits in the United States. To fill the void, other groups are setting up exchanges. One such exchange is the Chicago Climate Exchange.

The Chicago Climate Exchange (CCX), established in 2003, is a voluntary trading program. Members are legally bound to reduce emissions of all six major GHGs. Independent third party verification is provided by the Financial Industry Regulatory Authority (FINRA). Members make a voluntary commitment to meet an annual GHG emission reduction target. Credits are generated by members who reduce emissions below the targets while those who emit above the targets are bound to purchase CCX Carbon Financial Instrument®(CFI™) contracts. For further information go to www.chicagoclimatex.com.

How do you participate in one of the exchanges?

You must join an exchange in order to participate. Check websites for membership criteria, fees and commitments.

What is a cap and trade system?

A "cap and trade" system is different than buying and selling carbon credits on an exchange. Cap and Trade is a type of emissions regulation that uses the marketplace to find the lowest cost path to compliance by imposing a cap on the level of emissions. Cap and trade is a proven method for achieving environmental goals more quickly and at a lower overall cost than command and control regulation. Refer to the side bar for further information on emissions trading and cap and trade systems.

In a successful emissions trading system:

- Emission levels must be readily measured.

- Emission caps must be clearly defined and specified.

- Enforcement of standards and verification must be rigorous.

The Regional Greenhouse Gas Initiative (RGGI) is a regional cap and trade program geared specifically to address carbon dioxide emissions from power plants in the northeast and mid-Atlantic. States that are currently participating in RGGI include Connecticut, Delaware, Maine, Maryland, New Hampshire, New Jersey, New York, and Vermont. The states of Massachusetts, Pennsylvania, Rhode Island, the District of Columbia and the eastern Canadian provinces are observing the process. For more information, go to www.rggi.org.

In October, 2007, a partnership was formed among countries and regions actively pursuing carbon market development through mandatory cap and trade systems. International Carbon Action Partnership (ICAP) provides a forum within which to share experiences, research, and best practices on the process of designing cap and trade systems to reduce greenhouse gas emissions. The goal of ICAP is to help the different trading systems develop in a compatible manner in order to facilitate the transition to a global carbon market. Ten U.S. states are members: Arizona, California, Maine, Maryland, Massachusetts, New Jersey, New Mexico, New York, Oregon, and Washington. The other members of ICAP are: nine European Union countries, the European Commission, two Canadian provinces, New Zealand, and Norway. For more information, go to www.icap.com.

 **Cap and Trade Programs**

Using a cap and trade system to reduce pollutants is a proven method for achieving reductions more quickly and economically than imposing regulations (Emission Trading in the U.S., Pew Center on Global Climate Change). The Lead Trading Program for gasoline and Acid Rain Program for the electric industry are two examples.

The Lead Trading program was implemented in the 1980s to reduce the amount of lead in gasoline. Through 1982, the EPA enforced site-specific lead limits on refineries. In 1982, the rules of the program were amended to allow a refinery to use lead in its gasoline above the usual limit if it purchased the same amount of rights from other refineries that had successfully reduced their own lead content. Refineries were also allowed to "bank" credits for reductions. This change resulted in faster reductions in lead emissions and more efficient adoption of lead-reducing technologies by refineries.

The sulfur dioxide (SO_2) cap and trade program, part of the 1990 Clean Air Act, established a national cap of approximately 9 million tons of SO_2 emissions per year from electricity generating plants. The cap was implemented using tradable allowances issued to power producers – each allowance equaled the right to emit one ton of SO_2. Owners of fossil fuel-fired electrical generating units were required to give up an allowance for each ton of SO_2 emitted. This program reduced compliance costs and resulted in reduced emissions more quickly than anticipated.

Cap and trade systems work because the main objective of emissions trading – emission reductions at low cost – is achieved with a system that meets industry needs. Flexibility, banking of credits, clear definitions all make it possible for industry to economize. According to the Pew Center on Global Climate Change, emissions reduction goals have been enhanced by emissions trading (p.45).

For more information, go to www.pewclimate.org.

What are offset projects?

Exchanges may also grant credits for specific types of projects which reduce GHG emissions. These are known as "offset projects." Offsets are generated by non-regulated entities that create reductions. Generally, offset project generators do not have significant emissions of their own to reduce, but they are able to produce net reductions that are sold into the market place. The net reductions result from projects funded by the offset purchaser. Offset projects managers must document reductions and demonstrate that the project would not have gone forward without the funding provided by the offset purchaser.

Capturing carbon dioxide in forests and agricultural soils, increasing energy efficiency in buildings and factories, generating electricity from renewable sources such as wind or solar, and practicing no-till agriculture are all examples of offset projects.

Offset projects result in other environmental, social, and economic benefits in addition to reducing GHG emissions. These may include restoring degraded lands, improving water quality, creating jobs, and saving money.

Key Points

• Reductions in GHG emissions can be achieved through operational changes to reduce or avoid emissions or to sequester carbon.
• Carbon credits are a commodity, issued by markets to its members, for GHG reductions that are quantifiable, permanent, verified, and certified.
• Emissions trading involves the buying and selling of carbon credits.
• A cap and trade system is a regulatory structure that establishes a cap on emissions and leverages economics to drive reductions.

6

What is your next step?

You've read this far because you are interested in reducing your carbon footprint and curbing your contribution to global warming. Now it should be clear to you that reducing GHG emissions is a multi-step process.

First, you must define your boundaries and conduct a GHG inventory. This step will clarify which GHGs you emit and help you to identify ways to reduce emissions.

Second, you must set a goal for reducing your emissions and then develop and implement a plan to realize emission reductions.

Third, if your goal is to generate carbon credits for your reductions, you will need to join an exchange and conduct your inventory according to the exchange's rules including third party verification of your inventory and documented reductions.

Fourth, you can leverage all of your hard work and effort through brokering carbon credits for profit, for publicizing your sustainable approach to inspire others to do the same, or both.

There is little doubt that eventually regulations will be developed that *require* GHG emission tracking and reduction. Voluntary reporting and emission reduction now is a strategic decision which strengthens your credibility as a proactive citizen or environmentally sound organization or business. Whether your goal is to generate and market carbon credits, or to reduce your emissions because it is the right thing to do (or both), conducting a GHG inventory based on sound design principles is fundamental to making the necessary changes to reduce your carbon footprint.

Appendix A
Answer Key:
How much do you already understand about climate change and greenhouse gas inventories?

1. d

2. d

3. d

4. a

5. d

6. d

7. false

8. d

9. b

10. d

11. d

12. d

13. b

14. c

15. d

16. a

17. d

18. d

19. c

20. f

21. true

22. d

23. true

24. true

25. d

26. a

27. d

28. d

29. c

30. true

Appendix B
Key to Acronyms and
Abbreviations

CCX	Chicago Climate Exchange
EPA	Environmental Protection Agency
FINRA	Financial Industry Regulatory Authority
GHG	greenhouse gas
GHGs	greenhouse gases
GWP	global warming potential
ICAP	International Carbon Action Partnership
IPCC	International Panel on Climate Change
RGGI	Regional Greenhouse Gas Initiative
WBCSD	World Business Council for Sustainable Development
WRI	World Resources Institute

Appendix C
List of Figures, Tables and Side Bars

Appendix D
Sources and Further Reading

GHG Inventories and Emissions Trading

Chicago Climate Exchange http://www.chicagoclimatex.com
Greenhouse Gas Protocol Initiative www.ghgprotocol.org
International Carbon Action Partnership www.icap.com
Intergovernmental Panel on Climate Change www.ipcc.ch
Regional Greenhouse Gas Initiative (RGGI) www.rggi.org
Pew Center on Global Climate Change www.pewclimate.org
United Nations Framework Convention on Climate Change
www.unfccc.int
World Resources Institute www.wri.org

Non Profits

Clean Air—Cool Planet www.cleanair-coolplanet.org
The Climate Trust www.climatetrust.org
International Council for Local Government Initiatives www.iclei.org
Union of Concerned Scientists, www.ucsusa.org/global_warming
Database of state incentives for renewable energy, www.dsireusa.org

Government

United States Environmental Protection Agency, www.epa.gov
EPA Climate Leaders Program, www.epa.gov/climateleaders
National Renewable Energy Laboratory, www.nrel.gov
United States Department of Energy, www.eia.doe.gov

Publications

Brower, Michael and Warren Leon. *The Consumer's Guide to Effective Environmental Choices.* Three Rivers Press: New York. 1999.

Ellerman, A. Denny, Paul L. Joskow, David Harrison, Jr. *Emission Trading in the U.S., Pew Center on Global Climate Change.* May 2003.

Goodall, Chris. *How to live a low-carbon life: the individual's guide to stopping climate change.* EarthScan: Virginia. 2007.

World Resources Institute. *The Greenhouse Gas Protocol.* World Resources Institute and World Business Council for Sustainable Development.

About the Author

Judy Purman has worked in the environmental sector for the last twenty years helping companies of all sizes achieve their goals in environmental stewardship and compliance, program development and implementation, and employee training.

Ms. Purman has been a contributing writer for such publications as *The Journal of Environmental Horticulture* and *Biocycle Magazine.* She is an elected member of the Board of Directors for the United States Composting Council, a trade and professional organization involved in research, education, market expansion and promotion for the compost industry (www.compostingcouncil.org). She is also the appointed Chair of the Minnesota Pollution Control Agency's Environmental Innovations Advisory Council, which advises the MPCA Commissioner on policy, programs and legislation in pollution prevention, waste reduction, reuse, recycling and resource conservation.

Currently, Ms. Purman works with clients to develop comprehensive environmental stewardship programs designed to help push beyond the regulatory requirements and lessen their carbon footprint. She lives with her family in St. Paul, Minnesota. You can reach Judy at judy@thepurmangroup.com. For more information on sustainability, go to www.thepurmangroup.com.

Index

978-0-595-50141-0
0-595-50141-9